THE COMPLETE GUIDE
TO USING A
MULTIMETER

Mastering the Basics and
functions of a Multimeter

William M. Belville

Table of contents

Introduction

Welcome to your guide on using a multimeter! A multimeter is a vital instrument for every DIY enthusiast, electrician, engineer, or hobbyist. It is a flexible and powerful gadget that can measure a variety of electrical variables, such as voltage, current, resistance, and continuity. This guide is aimed to help you grasp the fundamentals of using a multimeter, including how to set it up, how to securely use it, and how to interpret the findings. With a multimeter, you can detect electrical faults, troubleshoot circuits, and take exact readings. This guide will equip you with the information you need to get the most out of your multimeter. We will start by examining the many varieties of multimeters, as well as their characteristics and capabilities. After that, we will explain how to set up and operate a multimeter, including how to pick the proper settings and interpret the findings. We will also explore how to troubleshoot and diagnose electrical faults using a multimeter, how to measure Resistance, voltage, Duty cycle, and so on. We will next explore the safety factors you must keep in mind while using a multimeter. By the conclusion of this book, you will have a full grasp of how to use a multimeter safely and efficiently. You will be able to utilize your multimeter to obtain exact measurements, troubleshoot circuits, and identify electrical faults. Now, let's get started!

Chapter 1:understanding The Multimeter

What is a Multimeter?

A Multimeter is an electronic instrument, an electronic technician, and engineer's extensively used piece of test equipment. A multimeter is usually used to measure the three fundamental electrical properties of voltage, current, and resistance. It may also be used to assess the continuity between two locations in an electrical circuit.

The multimeter has numerous features and operates as an ammeter, voltmeter, and ohmmeter.

It is a portable device featuring positive and negative indicator needles over a numeric LCD digital display. Multimeters may be used for checking batteries, home wiring, electric motors, and power supply.

The primary characteristics of the multimeter fundamentally consist of a display, power supply, probes, and controls.

How to use a Multimeter?

The purpose and functioning of a multimeter are similar for both analog and digital kinds. This instrument has two leads or probes notably red and black & three ports. The black color lead is utilized to connect to the common port, whilst the red color leads plug into various ports according to the demand.

Once the leads are connected, the knob may be turned ON in the middle of the instrument so that the proper function can be done for the particular component test. For instance, after the knob is located to 20V DC, then the multimeter will observe DC voltage up to 20V. To calculate low voltages, then adjust the knob of the multimeter to the 2V/200mV range.

To acquire a reading from the meter, you need to contact the end of each probe to the end of the terminals of components. Types of multimeter gadgets are very safe to employ on devices and circuits to provide a current or voltage that does not go beyond the maximum rating of the meter.

While measuring, we must be extremely careful so don't contact the bar ends of the metal in the tester when triggered or you will receive an electrical shock.

Functions of Multimeters

These devices are capable of varied readings dependent on the model. So basic forms of multimeter are usually used to measure amperage, resistance, and voltage, verify continuity and a whole circuit may be examined like the following:

- Resistance in Ohms
- Capacity in Farads
- The temperature can be measured in Fahrenheit or Celsius, as well as AC Voltage and Amperage.
- Inductance Henrys DC Voltage & Amperage
- Frequency in Hz
- Conductance in Siemens Decibels
- Duty Cycle

To certain kinds of multimeters, special sensors or accessories may be fitted for further measurements including acidity, light level, alkalinity, wind speed & relative humidity.

Types of Multimeter

There are numerous sorts of multimeters such as Analog, Digital, and Fluke multimeters.

Analog Multimeter

A VOM (Volt-Ohm-Milliammeter) or Analog Multimeter uses a moving coil meter and a pointer to show readings on the scale. The moving coil meter consists of a coil coiled around a drum positioned between two permanent magnets.

As current runs through the coil, the magnetic field is created in the coil which interacts with the magnetic field of the permanent magnets and the resulting force causes the pointer attached to the drum to deflect on the scale, indicating the current reading. It also includes springs attached to the drum which provides an opposing force to the motion of the drum to control the deflection of the pointer.

For the measurement of DC, the D'Arsonval movement mentioned above may be directly employed. However, the current to be measured should be smaller than the full-scale deflection current of the meter. For greater currents, the current divider rule is employed. Using multiple values of shunt resistors, the meter may also be used for multi-range current measurements. For current measurement, the instrument is to be coupled in series with the unknown current source.

For the measurement of DC voltage, a resistor is connected in series with the meter, and the meter resistance is taken into consideration so that the current traveling through the resistor is the same as the current running through the meter and the total reading displays the voltage reading. For voltage measurement, the instrument is to be linked in parallel with the unknown voltage source. For multi range measurement, various resistors of varying values may be used, which are linked in series with the meter.

For measurement of resistance, the unknown resistance is linked in series with the meter and across a battery, so that the current traveling through the meter is exactly proportional to the unknown resistance. For AC voltage or current measurement, the identical procedure is utilized, except for the fact that the AC parameter to be measured is first rectified and filtered to produce the DC parameter and the meter gives the RMS value of the AC signal.

The advantages of an Analog Multimeter are that it is affordable, doesn't need a battery, and can monitor changes in the readings. The two key parameters impacting the measurement are sensitivity and accuracy.

Sensitivity relates to the opposite of the full-scale deflection current and is measured in ohms per volt.

Digital Multimeters

We commonly use a multimeter as a digital multimeter (DMM). The DMM performs all operations from AC to DC other than analog. It has two probes positive and negative denoted with black and red color as shown in the image. The black probe links to COM JACK and the red probe is linked by user demand to measure ohm, volt, or amperes.

The jack labeled VΩ and the COM jack on the right of the photo are used for measuring voltages, and resistance, and for testing a diode. The two jacks are employed when an LCD is being measured (volts, ohms, amps, etc.). Overload protection avoids harm to the meter and the circuit and protects the user.

The Digital Multimeter comprises an LCD, a knob to choose different ranges of the three electrical properties, an internal circuitry consisting of signal conditioning circuitry, and an analog-to-digital converter. The PCB comprises concentric rings that are linked or detached depending on the position of the knob. Thus as the needed parameter and the range are chosen, the part of the PCB is triggered to do the relevant measurement.

To measure the resistance, current flows from a steady current source through the unknown resistor, and the voltage across the resistor is amplified and sent to an Analog to Digital Converter, and the resulting output in the form of resistance is shown on the digital display. To measure an unknown AC value, the voltage is first reduced to gain the correct range and then rectified to a DC signal, and the analog DC signal is supplied to an A/D converter to produce the display, which displays the RMS value of the AC signal.

Similarly to measure an AC or DC, the unknown input is first transformed into a voltage signal and then supplied to an analog-to-digital converter to acquire the required output(with rectification in case of an AC signal).

Highlights of a Digital Multimeter include its output display which quickly shows the measured value, remarkable accuracy, and ability to read both positive and negative values.

TYPES OF DIGITAL MULTIMETERS

Digital forms of multimeters are offered in three sorts.

Fluke Multimeter

The fluke digital multimeter may be developed with numerous collaborative functionalities. Generally, it comprises a big display and this device is used to measure the voltage alongside electrical resistance.

Some types of instruments are provided with extensive functions to measure humidity, duty cycle, pressure, frequency temperature, etc. The fluke multimeter is one of the most often renowned tools.

This form of the multimeter is frequently used for calibration efforts and employed to calibrate currents, volts & other electrical units.

The fluke multimeters are protected from the transient voltage. It is a compact portable gadget used to measure voltage, current, and test diodes. The multimeter features numerous pickers to choose the required function. The fluke MM automatically ranges to choose the most measures. This means the magnitude of the signal does not have to be known or estimated to take an accurate reading, it is immediately moved to the appropriate port for the intended measurement. The fuse is safeguarded to avoid harm if attached to the incorrect port.

Clamp Digital Multimeter

The clamp digital multimeter is used to measure the electrical flow. As the name implies, this multimeter has a feature namely a clamp that measures the amps whenever the probes sense the volts. The modification of power usage alternatively watts may be done simply by multiplying the measurement of voltage with the amps. This multimeter also offers an extra function that is various sorts of settings. The relevant feature is employed during the measuring.

This sort of multimeter has fixed equipment for measuring the current flow. This gadget dramatically alters from the fluke kind because, in the fluke multimeter, it employs a clamp to measure the flow of current. So, this gadget is normally advised for specialists only.

Autoranging Multimeter

The auto-ranging multimeter is an uncomplicated multimeter to utilize even though it is also the most priced of all types of digital multimeters. This multimeter contains a knob in the middle and has less position. So it doesn't switch automatically to measure. This instrument is appropriate for basic projects. For novices as well as electricians at home, this gadget is highly recommended. On average, it measures one component at a time.

TYPES OF MULTIMETER PROBES

A multimeter has several test probes and the primary purpose of these probes is to connect to the circuit under test. The most frequent forms of probes are retractable hook clips, pointed probes & crocodile clips.

Generally, a multimeter has two-color wires like black and red, known as leads or probes. One end of the probe is called a banana jack that is plugged into a multimeter, and the other end is known as the probe tip, used to test the circuit.

The red probe is used for +ve whilst the black probe is used for −Ve.

These probes have a probe tip on one end whilst the other end has banana plugs. Most of the multimeters feature fuses to shield them from the excessively high current.

When too much current is supplied via the multimeter, this fuse will block the flow of current to avoid damage. Some types of multimeters have fuses depending on the measurement of low current or high current and they dictate where you have to set the probes.

Working

Types of multimeters comprise two probes like red and black & two or three ports. From these, one of the ports is labeled.COM for ordinary which can be utilized for black probes whereas the other ports are labeled A for amps and mA/µA (milliamps/microamps). The fourth port is called VΩ used for ohms & volts.

Sometimes, this port is incorporated into the 3rd one, which is next named mAVΩ.

If the multimeter supports four ports, then the red probe may be placed into the VΩ port for measuring resistance as well as voltage. When the red probe is inserted into the mA port then the current can be calculated & linked into the A port then the current can be measured in amps. For instance, the port used to test a diode utilizing a multimeter is the VΩ port and the identical port may also be used to test a transistor

Chapter 2:Measuring Ac And Dc Voltage

What is Voltage?

The difference in electrical potential between two places is known as voltage or potential difference. It is measured in volts. Voltage may be either alternating AC or direct DC. The AC voltage switches polarity frequently, hence; it does not have distinct polarity. While the DC voltage stays constant with fixed polarity.

As the voltage is the potential difference between two places, measuring the voltage does not need you to open the circuit. You simply need readily accessible two locations in the circuit.

How to Measure Voltage with Digital and Analog Multimeters

Measuring AC Voltage with Digital Multimeter:

- To determine the voltage, the Multimeter has to be connected along with the voltage source, load, or any circuit.

- First, ensure sure the circuit that is to be studied is conveniently accessible for putting the probes.

- Turn the meter ON by toggling the ON/OFF button. Some switch ON by using the dial.

- Turn the dial to $\tilde{v}$ (V with a wavy symbol over the highest point of it). It refers to AC voltage.

- If there is the capability of range selection (some meters feature auto range mode for picking range depending on the reading), set the range to the greatest predicted value.

- If the voltage is uncertain, set the dial to the highest voltage range.

- Place the black probe into what is called the COM (common) socket. It is simple to distinguish in every meter.

- The red probe must be placed into its suitable socket. Other meters have a distinct socket for voltage whereas others have a composite socket for V-Amp-Ω. Insert the red probe into the socket with V on it.

- Place the black lead first on the lowest voltage location such as the ground.

- Place the red lead on the higher voltage spot.

- Note the reading in the multimeter.

- If the range is set to the largest voltage range, lower it step by step to reach maximum resolutions and receive an accurate result.

- Once the measurement is complete, remove the red probe first and then the black probe from the test locations.

- Remove both probes from the Multimeter and switch off the meter.

Note: Avoid handling the tips of the lead even if one of them is connected. Do not allow the points of the lead to contact one another. Be extremely cautious while working on the AC voltage of the mains since it might shock or electrocute a person if adequate safeguards are not followed.

Measuring AC Voltage with Analog Multimeter:

- Switch ON the multimeter

- Turn the dial to AC voltage.

- Using the selection knob on the dial, pick the most appropriate voltage range. It must not be less than the voltage being measured or overloading may harm the meter.

- If the voltage is unknown, set the knob at the largest voltage range accessible.

- Insert the black probe into the COM socket.

- Insert the red probe into the VΩ Or any socket that has a V sign on it.

- Various analog multimeters feature separate sockets for high and low voltages. Make careful you use the right socket.

- Connect the black lead directly to the low-voltage point.

- Next, put the red lead on the high-voltage point.

- Or just simply connect both leads across the component if uncertain about the voltage level.

- Ensure the deflection is greatest by reducing the voltage range using the adjuster. It boosts the precision of the reading.

- Note the reading on the scale. The reading must be done by seeing the meter at a straight angle to eliminate parallax mistakes.

- Based on the specified range (500), this measurement displays 220 v. since 500 is the full-scale deflection (FSD).

- When completed measuring, remove the red probe first and then the black probe.

- Turn off the meter and set the range at maximum to avoid any harm in case of fast reuse.

Note: AC voltage has no polarity. Therefore, it does not matter whether you swap the probes. The needle will always indicate the same degrees of deflection.

Measuring DC Voltage with Digital Multimeter:

- Ensure the circuit is conveniently accessible.

- Turn the meter ON by pressing the ON/OFF button or by rotating the knob on the dial from the off position.

- Turn the dial to DC voltage (V with straight light with 3 dots on top of it).

- If there is the capability of range selection (some meters feature auto range mode for picking range depending on the reading), set the range to the greatest predicted value. For instance, set the range to 20 when testing a 12 v circuit.

- If the voltage is uncertain, set the dial to the highest voltage range.

- Insert the rear probe into the COM (common) socket.

- The red probe must be placed into its suitable socket. Other meters have a distinct socket for voltage whereas others have a composite socket for V-Amp-Ω. Insert the red probe into the socket with V on it.

- Insert the black lead first on the lowermost or negative voltage point.

- Place the red lead on the higher voltage spot.

- Note the reading in the multimeter.

- If the range is set to the highest voltage range, lower it step by step to attain maximum resolution and receive an accurate result.

- Once the measurement is complete, remove the red probe first and then the black probe from the test locations.

- Remove both probes from the Multimeter and switch off the meter.

Take note that It is safe to perform tasks and touch a DC circuit at low voltage. However, it is preferable not to touch the tip of the leads when reading as it may impose mistakes in the measurement. Never touch the pointy ends of the leads altogether.

Measuring DC Voltage with Analog Multimeter:

- Turn the meter ON.

- Rotate the knob to DC voltage "VDC" or (V with a straight line having 3 dots with it).

- Set the range larger than the predicted value of the testing voltage.

- If the voltage is unknown, set the range to the largest available limit.

- Insert the black probe in the COM socket

- Insert the red probe in the VΩ socket ideally having VDC printed on it.

- Place the black probe on a negative or lower voltage point.

- Insert the red probe on a positive or increased voltage level.

- Reduce the voltage range to get maximum deflection to boost accuracy.

- Note the reading from the VDC scale (do not mix it with the VAC scale).

- When done, remove both probes red first and black second.

- Turn off the meter. Also, set the range to max to prevent harm during fast reuse.

Note: Mind the polarity when using an analog multimeter to measure DC voltage. It will not display any deflection when linked with opposite polarity. It may in certain situations cause harm to the meter.

Chapter 3:Measuring Ac And Dc Current

The pace in which electrons flow, i.e., current across a conductor is measured with an ammeter. To do the measurement of current using an ammeter, the circuit must be opened and then the meter is put in series or line with the circuit as shown in the figure.

This implies that an ammeter must be mounted in the path of the current flow where the current is being evaluated. These meters may be panel-mountable or portable. In this post, we are going to deal with a portable ammeter which is a component of the multimeter.

Both AC and DC currents may be measured with a multimeter by putting the meter in series with the circuit, involving the current is measured given the current in that circuit is restricted or regulated by a load or adequate amounts of resistance.

It needs to be mentioned that an ammeter is a low-resistance device and normally the impedance is less than 0.1 ohm.
If the meter has a connection in parallel to the supply mistakenly, this low resistance only reduces the current passage through the meter.

Suppose this meter with 0.1 ohms resistance is put across the 240 supply, the current flow would be roughly 2400 Amps (240/0.1 = 2400 A). This high current will lead to the depletion of the ammeter.

Thus, the ammeter must be connected in series or in line with the circuit in which the current is being measured. That's why ammeters are often termed in-line ammeters.

Current Measurement using Analog Multimeter

The operation of the analog ammeter is identical to that of the PMMC meter. In this, a resistor is put across the meter movement referred to as a shunt which restricts the amount of current that travels through the meter. Since the meter movement is coupled in parallel with the shunt, the voltage applied across the meter equals the voltage drop of the shunt.

Thus, the meter will show a full-scale reading when the rated current flows through the shunt as depicted in the picture below. And so the value of the shunt will vary based on the intended full-scale reading of the ammeter.

Many ammeters or multimeters are constructed to perform at more than one range, i.e., it enables several scales in one meter. This may be performed by connecting various shunts across the meter.

A rotary switch in series with these shunt resistors connects the required shunt across the meter depending on the range being measured as illustrated in the figure. Again the values of these shunts are determined dependent on the whole scale reading of the range (similar to that of single shunt meters).

For AC measurements, an analog multimeter includes a diode rectifier circuit that converts AC to corresponding DC. However, this diode has a definite switch

ON voltage which influences the low current readings (due to misrepresentation at low voltage side scale deflection).

This is one of the reasons to have limited ranges of AC currents in analog multimeters whereas certain meters may not monitor AC currents.

Whenever measuring the current via multimeter, one ought to take into account the following

1. Range selection knob for adjusting current knob.
2. DC or AC form of current
3. The expected range of current

4. Positioning of the red probe for DC and AC evaluations

Procedure to Measure DC Current with Analog Multimeter

- Insert red and black probes into the multimeter in their respective slots, depending on very high or very low current values. The 'mA' symbol slot indicates low current measurements and the 'A' symbol slot indicates high current measurements. In some meters, current values are directly written on their corresponding slots. The red probe must be inserted into these slots while the 'COM' sign slot constitutes the negative (or black) probe slot.

- Set range selection switch to DC measurement type of current and also pick the predicted range. It is usually preferable to ensure a maximum range for the measurement than that expected since we may also limit the range later if required. This eliminates the unwanted overload that might harm the meter.

- Switch off the power supply to the circuit in which the current is being monitored. And fracture the circuit to join the meter in series to the current route given load is connected to that circuit.

- Connect the red probe to the positive side (source) of the terminal and the back probe to the opposite side (load side or the end which is separated from the positive side) of the terminal. The meter will report a negative deflection if the probes are attached in a reverse way.

- Turn ON the power source and optimize the range of the multimeter by minimizing the selection switch steps for the maximum deflection of the pointer.

- Always remember to adjust the location of the probes after the current reading is done. And also flip the selection switch to the highest voltage position. This will prevent the risk of erroneously connecting the meter next time across the load while the multimeter is in ammeter mode. And so the harm to the meter is prevented.

Procedure to Measure AC Current with Analog Multimeter

Measuring the AC is comparable to measuring the DC as indicated before. There is not much difference between measuring AC and DC; still, some of the fundamental ways to perform the AC measurements are listed here.

- Place the red probe in the mA or A slot based on the range of AC that is being measured. Insert black probes into the COM slot.

- Set the range selector switch to AC type of current and set the maximum range for the current measurement.

- Switch off the power supply to the circuit and make sure to separate the route of current (i.e., phase of the circuit) at which current is to be measured to connect the meter in line with the circuit.

- Put the red probe to the supply side of the phase terminal and the back probe to the reverse side of the phase terminal.

- Turn ON the power source and optimize the range of the multimeter by minimizing the selection switch steps for the maximum deflection of the pointer.

Analog meters are equipped with adjustment screws to fix the needle position to zero.

So make sure to verify that the needle is positioned at zero position when monitoring is going to measure. If not, modify it appropriately.

Current Measurement using Digital Multimeter

Digital multimeters are generally used as portable measuring equipment which has the advanced qualities of analog meters which include auto polarity, auto zeroing, auto range, and automatic shut-off.

This multimeter is connected in line with the wire or component by breaking the circuit sections to measure the current. Most of the DMMs include several ports for a wide variety of current measurements.

To measure the current (AC or DC) using a DMM, the internal circuit first changes the current at the input to a voltage to utilize by the ADC. This is addressed by a series of switching resistors sometimes called shunts. According to Ohm's rule, these shunts define the voltages proportional to the input currents to be measured.

AC is monitored in the same way as shunts except that voltage across the shunt is sent towards AC to the DC rectifier circuit before delivering it to ADC.

DC/AC Current Measurement using Digital Multimeter

- Plug the negative probe lead (black color probe) into the COM jack and the positively charged probe lead (red color probe) to either very low (mA or μA) or very high current range (A) jack following the greater magnitude of the current being measured.

- Suppose the meter contains 200mA and 10 A jacks, connect the red probe to the 200mA jack for the maximum of 200mA current measurement, on the other hand, link the red probe to the 10 A jack for the optimum of 10A current assessment.

- Set the kind of current AC or DC.

- Set the range selection switch at the ideal range that offers the greatest sensitivity or just pick a high range so that later we may reduce the steps if required (Some DMMs are of auto range meters thus no need for choosing the range).

- Turn OFF the circuit and interrupt it at the point when the reading is to be taken.

- Put a red probe to the more positive side as well as a black test to lead to the negative end of the circuit.

- Turn ON the power source and change the range of current to a closer digital form.

- If the meter reads 'OL', it is indicating an over-range situation and therefore the selector switch, i.e., the range has to be modified suitably.

- If the meter is positioned in 200mA (probe connection), indicating that the maximum input current permitted by the meter is 200mA. If the current is exceeded, the fuse of the meter will be damaged. Also, when the meter is put in 10A, the maximum current is 10A for which no fuse safety is given.

Cautions

Never leave the multimeter in the ammeter position after the current measurement is made.

Don't test greater currents than that of the maximum current detected by the multimeter in their respective ranges, i.e., mA as well as A range.

Chapter 4:Measuring Resistance

How to Measure Resistance with Digital & Analog Multimeter?

Measuring Resistance using a Multimeter? (DMM – Analog Meter)

Measuring resistance much like voltage and current is a crucial aspect of diagnosing any component. It indicates the state of the components. The resistance measurement is also used to check for open or closed circuits. Last but not least you may examine how precise the resistors are as they are also color-coded.

What is Resistance?

Resistance is the antagonism to the passage of the current. The equipment that is used for measuring the resistance is called Ohmmeter. Ohmmeter has a voltage across its two terminals that flow the current through the component being evaluated. If the resistance is high it signifies that little current will flow. If the resistance is low that suggests the current flowing will be high. Based on the quantity of current flowing, it calculates the resistance.

We may also employ the resistance data to detect a short circuit or a damaged circuit. In this multimeter lesson, we will measure the resistance using DMM and an analog multimeter with step-by-step guidance.

Before you begin to follow the procedures, you may know the fundamental difference between AC and DC resistance.

Measuring Resistance with Digital Multimeter:

- Toggle off the power supply to the circuit.

- If there is a capacitor on board, remove it first.

- Isolate the component whose resistance has to be tested. If feasible remove it from the circuit to eliminate any parallel routes that may interfere with the overall resistance.

- Switch ON the multimeter.

- Turn the selection knob to the resistance Ω

- Select an appropriate range somewhat higher than the predicted resistance value for optimal accuracy. If it is unknown, pick the higher settings. It can be brought back down later.

- Insert the black probe in the COM (common) socket.

- Insert the red probe in the Ω Most DMMs feature a common socket used for Ω, V, and continuity. Use the socket that has an Ω emblem on it.

- Connect the leads across the component.

- Note the reading. Change the range to the lowest possible level to achieve an accurate reading.

- When done, remove the probes and switch the selection knob into the voltage mode to avoid harm by accidently connecting to high voltage.

Take note: Do not measure the resistance of a circuit with power still ON. Look out for capacitors in a circuit before testing a component. The components in parallel also impact the equivalent resistance. ensure sure the component being tested does not have any other component in parallel. Do not touch the tip of the leads during measuring, it will impose mistakes in the reading.

Measuring Resistance with Analog Multimeter:

Analog multimeter follows the same approach. However, it has a small calibration to complete when measuring resistance.

- As usual, cut off the power supply to the circuit first and discharge if there is any capacitor.

- The component being examined can't have any component in parallel.If feasible, remove the component from the circuit.

- Switch ON the Analog multimeter.

- Turn the selection knob to the resistance Ω

- Select an appropriate range somewhat higher than the predicted resistance value for optimal accuracy. It may be changed back afterward.

Note: The analog multimeter resistance range contains multiplication factors. For example, x1, x10, and x100 are distinct ranges indicating the scale value multiplied by the factor to reach the actual reading.

- Plug the black probe into the COM (common) socket.

- Immerse the red probe in the Ω Some meters belong to the Ω socket with voltage. Use the socket bearing the Ω sign on it.

- Calibrate or modify the zero of the meter by connecting both probes and moving the zero adjustment knob to display full-scale deflection i.e. o ohms.

- Connect the leads across the component.

- Note the reading. Adjust the range of the meter to indicate the greatest potential deflection to gain optimum accuracy.

- If the range is x1, then this measurement is 100 ohms. If the spectrum is x10, the reading is 1000 ohms. If the range is x100, the reading is 10,000 ohms.

- After finishing, remove the probes and choose the voltage measurement mode to prevent unintentionally attaching it to voltage.

Factor Affecting Resistance Readings

The resistance may be modified by several variables. Therefore, in measuring resistance the following elements must be considered:

The component in a circuit: If the component is within a circuit, its resistance may be altered by any other components in parallel.

Power through the circuit: if there is power provided to the circuit or any charged capacitor, it will alter the results as the ohmmeter operates based on the current flowing through the meter.

Diode in a circuit: If there is a diode in a circuit, the resistance of the circuit will fluctuate if the probes are swapped with each other. That is because the diode does not allow current in one direction.

Fingers contacting the leads: if your fingers are touching the leads, it will influence the reading owing to the leaking of some electricity via your body. Do not touch the tip of the leads when measuring resistance.

Temperature: Most components' temperature rise when current travels through them. It is preferable not to test the resistance while they are hot since the temperature influences the resistance.

Chapter 5:Continuity Test

Ways To Perform a Continuity Test for Electrical Components Using a Multimeter

What is a Continuity Test?

A continuity Test is the examination of an electrical circuit to see whether the current may travel through it (known as a close or full circuit).

In a continuity test, a small voltage is delivered to the two points of the circuit that need to be inspected. The current flow between these two places defines whether it's an open or closed circuit. Usually, there is a buzzer or led in series (inside continuity meter) to determine if the current goes through it or not.

A closed circuit offers a closed channel for the current flow & an open circuit does not allow the current flow. These circuits may be recognized using the continuity test.

Why Do We Use Continuity Tests?

Continuity test is a highly crucial test in troubleshooting any circuit. Various applications of continuity testing are:

- To inspect the wire connection within the circuit. These wires may be damaged.
- It is used for Identifying damaged components.

- It is also used for testing the quality of soldering.
- It is used for identifying a particular cable or electrical connection.

Procedure Of Continuity Test

There are essentially two ways for testing the continuity of a circuit using a multimeter.

The first option is to utilize the continuity mode in the multimeter, which is expressly developed for this purpose.

The second technique is to utilize the Ohmmeter.

USING CONTINUITY MODE

The procedures for continuity test using continuity mode are mentioned below:

- De-energize the circuit, if it has any electrical input.

- Set the dial of the multimeter in continuity mode (continuity mode is denoted by the sign of sound)

- Plug the black probes into the COM port.

- Connect the red probe to the V, Ω port.

- Now contact the probes with each other. If the meter beeps or provides a reading of 0 that implies the meter works well.

- Now attach the probes to both ends of the component or wire that you wish to test.

- If the meter reads 0 and beeps, it signifies the passage is complete (close) or the component enables the flow of current.

CONTINUITY TEST MODE

If the meter does not beep & display 1 or OL, it implies the route is broken (open) or the component does not allow the flow of current.

The continuity is non-directional, it does not matter which probe should be coupled to which side. The outcome is usually the same except in rare circumstances like diodes which allow the flow in just one direction.

Using Ohm-Meter

An Ohmmeter may also be used to assess the circuit whether it is a closed or open circuit, which is the primary objective of a continuity test.

Steps for continuity test using an ohmmeter:

- First de-energize the circuit, if it has any power supply.

- Set the knob of the multimeter to resistance mode Ω. If it has many ranges, set the dial to the least range.

- Push the black probe into the COM socket of the multimeter.

- Insert the red probe into the V, Ω socket.

- Connect the probes to both ends of the wire or component you wish to test.

- If the meter reads 0 Ohm or near 0 Ohm, the route is complete and close.

- If the meter shows 1 or OL, the wire connection is broken (open).

CONTINUITY TEST FOR CAPACITOR

You may examine a capacitor employing the continuity test.

- Disconnect the capacitor if it is in a circuit.
- Discharge it gently if charged.

Using Continuity Mode

- Set the Multimeter in continuous mode & insert the black & red probes as instructed previously.
- Place the red and black probes of the multimeter across the positive & negative terminals of the capacitor accordingly.
- If the capacitor is good, the readout should start at 'o' while the capacitor is charging from the multimeter. The

reading will grow & finally reach infinity or OL, which implies that the capacitor got completely charged & open.

- If a capacitor is destroyed, a multi-meter will either indicate an extremely low value (short) or infinite OL (open).

Using Resistance Mode

- Configure the dial of the multimeter in resistance mode.

- Attach the red probe on the positive terminal and the black probe on the negative input of the capacitor
- If the resistance begins growing from 0 Ohm to infinity, the capacitor is OK.

Because it was charging in the beginning.

- If the meter displays extremely high resistance initially even though it was discharged, the capacitor is damaged (open).

- If the results reveal extremely little resistance, the capacitor is short.

Chapter 6:Capacitance And Diode Testing

How to evaluate Diodes with a Digital Multimeter

Digital multimeters can test diodes employing one of two methods:

Diode Test mode: nearly usually the best technique.

Resistance mode: generally utilized only if a multimeter is not equipped with a Diode Test mode.

Note: In some instances, it may be essential to remove one end of the diode from the circuit to test the diode.

Things to know about the Resistance mode while testing diodes:

- Does not always show if a diode is good or poor.

- Should not be taken while a diode is connected in a circuit as it might create a misleading reading.

- CAN be used to verify a diode is defective in a particular application if a Diode Test shows a diode is bad.

A diode is best tested by measuring the voltage drop across the diode when it is forward-biased. A forward-biased diode operates as a closed switch, letting current pass.

Steps for using a multimeter in the diode test mode

A multimeter's Diode Test mode provides a modest voltage between test leads. The multimeter then shows the voltage drop when the test leads are connected across a diode while forward-biased. The Diode Test process is done as follows:

- Make assured a) all power to the circuit is OFF and b) no voltage exists at the diode. Voltage may be present in the circuit owing to charged capacitors. If so, the capacitors need to be emptied. Set the multimeter to measure ac or dc voltage as desired.

- Push the dial (rotary switch) to Diode Test mode. It may share a location on the dial with another function.

- Connect the test leads to the diode. Record the measurement shown.

- Reverse the test leads. Record the measurement shown.

Diode test analysis

- A reasonable forward-based diode has a voltage drop ranging from 0.5 to 0.8 volts for the most extensively used silicon diodes. Some germanium diodes have a voltage loss ranging from 0.2 to 0.3 V.

- The multimeter indicates OL when a good diode is reverse-biased. The OL value shows the diode is acting as an open switch.

- A faulty (opened) diode does not enable current to flow in any direction. A multimeter will indicate OL in both directions when the diode is opened.

- A shorted diode has the same voltage drop measurement (about 0.4 V) in both directions.

A multimeter set to the Resistance mode (Ω) may be used as an additional diode test or, as noted previously if a multimeter does not include the Diode Test mode.

A diode is forward-biased when the positive (red) test lead is on the anode and the negative (black) test lead is on the cathode.

- The forward-biased resistance of a good diode should vary from 1000 Ω to 10 MΩ.

- The resistance reading is high when the diode is forward-biased due to current from the multimeter flowing through the diode, generating the high-resistance measurement essential for testing.

A diode has an opposite bias whenever the positive (red) test lead is on the cathode and the negative (black) test lead is on the anode.

- The reverse-biased resistance of a good diode appears OL on a multimeter. The diode is faulty if readings are the same in both directions.

The resistance mode technique is executed as follows:

1. Make assured a) all power to the circuit is OFF and b) no voltage exists

at the diode. Voltage may be present in the circuit owing to charged capacitors. If so, the capacitors need to be emptied. Set the multimeter to measure ac or dc voltage as desired.

2. Turn the dial to Resistance mode (Ω). It may share a location on the dial with another function.

3. Connect the test leads to the diode once it has been removed from the circuit. Record the measurement shown.

4. Reverse the test leads. Record the measurement shown.

5. For optimal results when using the Resistance mode to test diodes,

compare the values obtained with a known good diode.

How to measure capacitance

A multimeter measures capacitance by charging a capacitor with a set current, measuring the resulting voltage, then calculating the capacitance.

Warning: A good capacitor holds an electrical charge and may stay active when power is disconnected. Before handling it or taking a measurement, a) turn all power OFF, b) use your multimeter to validate that power is OFF and c) gently discharge the capacitor by attaching a resistor across the leads (as described in the following paragraph). Remember to use adequate personal protection equipment.

Steps for measuring capacitance using a digital multimeter

To safely discharge a capacitor: After power is gone, put a 20,000 Ω, 5-watt resistor across the capacitor terminals for five seconds. Use your multimeter to ensure the capacitor has entirely depleted.

1. Use your digital multimeter (DMM) to confirm all power to the circuit is OFF. If the capacitor is used in an ac circuit, configure the multimeter to measure ac voltage. If is utilized in a dc circuit,

configure the DMM to measure dc voltage.

2. Visually check the capacitor. If leaks, fractures, bulges, or other indicators of degradation are obvious, replace the capacitor.

3. Turn the dial to the Capacitance Measurement mode. The symbol typically occupies a space on the dial with another function. In addition to the dial adjustment, a function button normally has to be pushed to trigger a measurement. reach out to your multimeter's owner's manual for directions.

4. For a precise estimation, the capacitor will need to be disconnected from the

circuit. Discharge the capacitor as specified in the warning above.

Note: Some multimeters feature a Relative (REL) mode. When measuring low capacitance values, the Relative mode may be used to eliminate the capacitance of the test leads.
To set a multimeter in Relative mode for capacitance, leave the test leads open and push the REL button. This eliminates the residual capacitance value of the test leads.

5. Connect the test leads to the capacitor terminals. Maintain test leads connecting for a few seconds to allow the multimeter to recognize the proper range.

6. Read the measurement indicated. If the capacitance value is within the measuring range, the multimeter will show the capacitor's value. It will indicate OL if a) the capacitance value is greater than the measuring range or b) the capacitor is defective.

Capacitance measurement summary
Repairing single-phase motors is one of the more practical uses of a digital multimeter's Capacitance Function.

A capacitor-start, single-phase motor that fails to start is an indication of a defective capacitor. Such motors will continue to operate once functioning, making troubleshooting tough. Failure of the hard-start capacitor for HVAC compressors is an excellent illustration of this issue. The compressor motor may start but rapidly overheat resulting in a breaker trip.

Single-phase motors with such difficulties and loud single-phase motors with capacitors need a multimeter to test correctly operating capacitors. Almost all motor capacitors will have the microfarad value indicated on the capacitor.

Three-phase power factor correction capacitors are normally fuse-protected. Should one or more of these capacitors fail, system inefficiencies will develop, utility costs will most likely increase and unintended equipment tripping may occur. Should a capacitor fuse break, the suspected defective capacitor microfarad value must be tested and confirmed if it falls within the range specified on the capacitor.

Some other issues affecting capacitance are worth knowing:

- Capacitors have a limited life and are frequently the cause of malfunction.

- Faulty capacitors may have a short circuit, or an open circuit or may physically degrade to the point of failure.

- When a capacitor short circuits, a fuse may explode or other components may be destroyed.

- When a capacitor opens or deteriorates, the circuit or circuit components may not perform.

- Deterioration may also modify the capacitance value of a capacitor, which might create difficulties.

Chapter 7:How To Test A Capacitor Using Digital And Analog Multimeters

In most electrical and electronics troubleshooting and repairing activities, we confront a typical issue with capacitors where we wish to know how to test and check a capacitor. Is it good, terrible (dead), short, or open?

Here, we may inspect a capacitor using an analog (AVO meter i.e. Ampere, Voltage, Ohm meter) as well as a digital multimeter either the capacitor is in excellent condition, or should we replace it with a brand new one

Note: To discover the value of Capacitance, you need an analog or digital multimeter with capacitance measurement capabilities.

Below are techniques to examine & test whether a Capacitor is Good, Defective, Open, Dead, or Short.

METHOD 1

Test a Capacitor employing Digital Multimeter – Resistance Mode

To test a capacitor using DMM (Digital Multimeter) in the Resistance "Ω" or Ohm mode, apply the steps described below.

- Make sure the capacitor is entirely depleted.

- Place the meter on the Ohmic range (Set it for a minimum of 1000 Ohm = 1kΩ).

- Attach the multimeter probes to each of the capacitor ports (Negative to Negative and Positive to Positive).

- A digital multimeter will display some figures for a second. Note the reading.

- And then instantly it will revert to the OL (Open Line) or infinity "∞". Every try of Step 2 will produce the same outcome as seen in Steps 4 and 5. It signifies that Capacitor is in Good Condition.

- If there is no Transition, then Capacitor is useless.

METHOD 2

Inspect a Capacitor using Analog Multimeter – Ohm Mode

To inspect a capacitor using AVO (Ampere, Volt, Ohm Meter) in the Resistance "Ω" or Ohm mode, complete the following methods.

- Make sure the suspicious capacitor is depleted.

- Take an AVO meter.

- Rotate the knob on the analog meter to pick the resistance "OHM" mode (Always, select the greater range of Ohms).

- Connect the Meter leads to the capacitor terminals. (COM to the "-Ve" and Positive to the "+Ve) terminals).

- Note the reading and compare it with the following findings.

- **Short Capacitors:** Shorted Capacitors will exhibit extremely little resistance.

- **Open Capacitors:** An Open Capacitor doesn't show any motion (Deflection) on the OHM meter scale.

- **Good Capacitors:** Initially, it will exhibit low resistance, and then gradually climbs toward the infinite. It signifies that the capacitor is in excellent condition.

METHOD 3

Measuring Capacitor employing Multimeter in the Capacitance Mode

Note: Testing a capacitor in the capacitance mode can only be conducted if the analog or digital multimeter has the farad "Farad" of Capacitance "C" functions. The function of capacitance mode in a multimeter may also be utilized to test the small capacitors. To do this, rotate the knob of the multimeter to the capacitance mode and abide by the subsequent easy steps.

- Make sure the capacitor is entirely depleted.

- Remove the capacitors from the circuit board.

- Now Select Capacitance "C" on the multimeter.

- Now connect the capacitor terminal to the multimeter leads. (Red to Positive and Black to Negative).

- If the reading is close to the real value of the capacitor (i.e. the written value on the Capacitor container box).

- Then the capacitor is in excellent condition. (Note that the reading may be smaller than the actual value of the capacitor (the rated value of the capacitor owing to the tolerance in ±10 or ±20).

- If you read a substantially lower capacitance or none at all, then the capacitor is dead and you need to swap it with a new one for correct functioning.

METHOD 4

Testing a Capacitor By Simple Voltmeter

To perform this approach on polar and nonpolar capacitors, you must know the value of the nominal voltage of capacitors.

The amount of voltage is already written on the nameplate of electrolytic capacitors. While there are unique codes written on ceramic and SMD capacitors. You may follow this tutorial which demonstrates how to read and discover the value of ceramic and non-polarized capacitors with relevant codes written on them.

Also, you may utilize the DC Voltage "V" or Volt Mode in the digital or analog multimeter to make this test.

- Make sure you disconnect a single lead (no concerns whether Positive (long) or negative (short)) of the capacitor from the circuit (You may detach as well if required)

- Check the capacitor voltage rating written on it (As demonstrated in our following example where the voltage = 16V)

- Then charge this capacitor for a few seconds to the rate rather than to the exact amount but less than that i.e. charge a 16V capacitor using a 9V battery. If the value of the battery voltage is larger than the nominal voltage of the capacitor, it will damage or burst the capacitor.) voltage. Make careful you connect the positive (red) lead of the voltage source to the positive lead (long) of the capacitor and negative to negative. If you are not sure or unable to discover the right leads, here is the lesson on how to find the negative and positive terminal of a capacitor.

- Set the value of the voltmeter to the DC voltage and connect the Capacitor to the voltmeter by connecting the positive wire of the battery to the positive lead of the capacitor and negative to negative. You may use a digital or analog multimeter when choosing the DC voltage range for the same reason.

- Note the first voltage measurement in the voltmeter. If it is near the given voltage you gave to the capacitor, the Capacitor is in Good condition. If it displays considerably less reading, Capacitor is dead then. notice that the voltmeter will display the reading for a very brief period as the capacitor will discharge its stored volts in the voltmeter.

Note: The value of the capacitor voltage should be smaller than the battery voltage. In addition, it will explode or destroy the capacitor.

METHOD 5

Test the Capacitor via Continuity Test Mode

In the DMM and AVO meter, the continuity test mode could have been employed whether the capacitor is good, open, or short. To do so, observe the straightforward procedures below.

- Turn off the electric supply and unplug the capacitor from the circuit board.

- Fully discharge the capacitor using a resistor.

- Rotate the knob and place the multimeter in continuity test mode.

- Make contact with the positive (RED) probe of the multimeter to the Anode (+) and the Common (Black) probe to the Cathode (-) terminal of the capacitor.

- If the multimeter displays an indication of correct continuity (beep sound or LED light) then abruptly stops and indicates an OL (open line). It signifies the capacitor is in excellent condition.
- If the multimeter doesn't exhibit a continuity indicator with a beep or light, that signifies the capacitor is open.

- If the multimeter LED lights ON and generates a persistent beep sound, it implies the capacitor is short and it should be replaced with a new one.

Chapter 8: Duty Cycle And Frequency

How to Measure Duty Cycle with a Digital Multimeter

Set the digital multimeter (DMM) to measure frequency. The steps might vary by the meter. Usually, a multimeter's dial will be switched to dc V, and the Hz button is pushed.

The DMM is poised to measure the duty cycle once a percent sign (%) emerges on the right side of the multimeter's display.

- First, insert the black test lead in the COM jack.

- Then put the red lead into the V Ω jack. When done, remove the leads in reverse order: red first, then black.

- Insert your test leads into the circuit to be examined.

- Read the measurement in the display. A positive sign (+) denotes POSITIVE time percent voltage measurement. A negative sign (-) denotes NEGATIVE time percent voltage measurement.

Note: A positive reading normally reflects a circuit's ON time and a negative reading its OFF time. On occasion, a negative component of the signal might provide an ON signal.

Push the beeper button to choose between POSITIVE time and NEGATIVE time % voltage measurement. Note: The button used differs per digital multimeter. Refer to your model's user manual for precise instructions.

Duty cycle fundamentals

The duty cycle is the proportion of time a load or circuit is ON to the time a load or circuit is OFF. A load that is switched ON and OFF multiple times per second has a duty cycle.

Why do this?

Various loads are swiftly cycled on and off by a fast-acting electronic transform to accurately manage output power at the load. Lamp illumination, heating element outcomes, and magnetic power of a coil may be controlled by the duty cycle.

The duty cycle is quantified in the percentage of ON time. Example: A 60% duty cycle is a signal that is on 60% of the time and off 40% of the time. Another approach to quantify the duty cycle is dwell, computed in degrees instead of %.

During the measuring duty cycle, a digital multimeter indicates the amount of time the signal being measured is above or below a specified trigger level - the fixed level at which the multimeter counter is activated to record frequency. The slope is the waveform edge on which the trigger level is chosen.

The % of time above the trigger level is indicated if the positive trigger slope is chosen. On the other hand, the proportion of the amount of time underneath the trigger level is given if the negative trigger curve is selected. The slope chosen is denoted by a positive (+) or negative (-) sign on the display. Most multimeters default to show the positive trigger slope; the negative trigger slope is generally set by pushing an extra button.

How to Measure Frequency with a Multimeter

Frequency is the number of cycles performed in one second. Many sorts of multimeters can measure frequency. Alternating current and other electrical signals contain frequency that impacts the functioning of a device. By applying a multimeter, we can measure numerous values such as voltage, current, resistance, capacitance, frequency temperature, and continuity, as well as test electrical and electronic components including resistors, capacitors, diodes, transistors and cables & wires etc.

In this post, we are going to investigate how a multimeter measures the frequency and what are the elements that impact its reading.

Working Principle

A digital multimeter that can measure frequency features a peak-detection circuit. The meter measures the time between the two successive crests (peak of waveform) using the peak-detection circuit. it recognizes the peak of the input waveform and begins the timer. When the next peak of the waveform is identified, it pauses the timer. The meter estimates the frequency using the time between the two crests of the waveform.

Any digital multimeter that can measure frequency has "Hz" inscribed anywhere on the dial and onto the ports where probes are placed. It may also share a location on the dial with "VAC" or "V~".

There are two techniques for detecting frequency on a multimeter. If your multimeter has a specific area on the dial, then follow this procedure.

METHOD 1

- Turn the meter ON by toggling the ON/OFF button.

- Turn the dial to "Hz", it shares space on the dial with any other function such as "VAC or V~".

- Pressing the "shift" button accesses the second option and begins monitoring frequency. "Hz" displays on the display to signify the meter has shifted to frequency measurement.

METHOD 2

Some multimeters feature a designated place for frequency measurement on the dial that has "Hz" printed on it.

- Plug the black probe first through the "COM" port.

- Then place the red probe into the port with "Hz" inscribed on it.

- Connect the black lead first and then the red lead to the location of measurement.

- Note the reading from the display.

- If your multimeter has several ranges, limit the range to achieve an accurate measurement. Most multimeters offer an "auto range" button to pick the right range depending on the reading.

- When done, remove the red lead first and then the black lead.

- Remove the black and red lead from the meter ports.
- Set off the meter or set the dial to "voltage measurement" to prevent any possible harm in case of hasty reuse.

Problems Incurred during Frequency Measurement

Various difficulties might alter the frequency reading of a multimeter. We can minimize some of them to achieve an accurate reading.

Range of the Meter

The datasheet of a multimeter states the lowest and highest frequency the meter can correctly measure. If the input frequency falls below the range, the multimeter may report a reading near the real measurement but not precise enough. The same thing will happen at a higher frequency above the range. The meters may not keep up with the real frequency and display lower values or exhibit "OL" overload.

Therefore, it is vital to know the range of the meter and the estimated frequency of the input signal.

Distortion in Input Signal

If the input signal includes frequency distortion, it might impact the reading of the multimeter and produce ambiguity in the measurement. The reading may also vary. The signal may be filtered from noise by employing a low-pass filter.

Signal Radiation

Sometimes, the multimeter may pick up the frequency measurement without probes contacting the line. it may arise owing to the unshielded wires that function as antennas to emit the EMI (Electromagnetic interference). The meter takes up the signal, amplifies and measures It, and shows the reading. It may or may not be accurate. Therefore, it is better to physically attach the probes to the wire.

Why do We Measure Frequency?

Measuring frequency is crucial because circuits and devices are built to work at specified frequencies. They either run at constant frequency or variable frequency where the output depends on it.

One such example is an AC electric motor whose speed is exactly proportional to the frequency of the main supply. A motor or transformer intended to work at 50 Hz will run at a greater speed if connected to a 60 Hz source. Similarly, a 60 Hz motor and transformer will be slower if it operates on a 50 Hz supply. Here is an interesting topic for you :) Is it Possible to Operate a 50Hz Transformer on 5 Hz or 500 Hz Frequency?

Good to Know: Frequency can't be measured with an Analog Multimeter. Only specialist digital multimeters with particular capabilities (such dedicated Auto Hz button or separate Com terminal) may be used to measure frequency in various ranges even up to 100kHz or more.

Chapter 9:Measuring Temperature

Not all multimeters can detect temperature. Before you begin, verify whether your multimeter can detect temperature. If your meter can measure temperature, you will see a thermometer sign on the multimeter dial.

To measure temperature with your multimeter, you are not going to utilize the normal black and red leads. Instead, you are going to utilize a thermocouple and a thermistor.

A thermocouple is a rectangular brick with two metal prongs. One prong will contain a positive (+) sign. The other is a negative (-) sign. The thermocouple has a sheathed wire running from it. We use this cable to monitor the temperature.

The negative prong of the thermocouple will be hooked into the COM port. The positive prong of the thermocouple will be inserted into the temperature port. The emblem for the temperature connector is a thermometer. It is generally the same port used for voltage.

The thermocouple has a wire running out of it with a thermistor at the end of the wire. A thermistor is a resistor whose resistance is dependent on temperature. Based on the resistance of the thermistor, the multimeter can read the temperature.

How to measure temperature with a multimeter

Let's go through the procedure of monitoring temperature.

- Start by checking that the negative end of your thermocouple is inserted into the COM port.

- The positive end of the thermocouple should be put into the temperature port.

- Rotate your dial until it points toward the temperature sign. You may need to utilize the function key to access the temperature measurement.

- When your meter is set to a temperature, you will see an "F" or "C" on the screen. These stand for Fahrenheit and Celsius.

- Take note that there are two metrics for measuring temperature: Celsius, and Fahrenheit

You will need to configure your multimeter to measure the right units.

To switch your multimeter between Fahrenheit and Celsius, you will need to press the orange button. While your dial is set to a temperature, push the orange button to change units. The unit will be on display. For example, the units on the display flip from "F" to "C" when you hit the orange button.

Once you have checked that your meter is in the right setting, grasp the wire coming out of the thermocouple. At the end of this cable is a thermistor that the meter will use to monitor the temperature. Right now, the meter should be indicating the air temperature.

To measure the temperature of an item, push the tip of the thermistor onto the object. The meter's display will slowly begin to alter when the resistor warms up or cools down. Wait for the temperature to be a consistent value. That is the temperature of the thing.

Chapter 10:Multimeter Safety Precautions

While using a multimeter sounds very basic and straightforward, there are a handful of things you need to worry about, particularly if you aren't as experienced. Working with electricity should always be viewed as potentially harmful, regardless of the low numbers.

Many individuals assume they aren't in imminent danger owing to low voltages or amperage, but you can rest confident that there is a certain risk associated, particularly if you don't operate the unit correctly and by the guidelines.

Safety Tips When Using a Multimeter

Inspection of the Unit

Before you even start utilizing certain equipment, be sure to take the time and check it out first. Look for indicators of physical deterioration, cracks, leaks, or anything that might endanger the accuracy of the findings as well as your safety.

One should never presume that the device is operating correctly.

In a nutshell, you should always check the multimeter on a voltage that is an established source before working with a high-current circuit. This phase enables you to validate that the device is operating as expected. In case you obtain incorrect readings from a recognized source, it is a good sign that something is amiss. Needless to say, don't utilize a malfunctioning model on an unknown supply, especially if excessive voltages are in worry.

Also, while not in use, be sure to keep the multimeter in its protective case together with the test probes.

Inspection of the Probes

The functioning of the leads is as vital as the functionality of the multimeter. If one of the two probes (or both) is malfunctioning, there is a substantial danger of damage involved, particularly at high voltages. That's why you must test the probes before utilizing them on an unknown source.

The first step is to disconnect the probes and assess for physical damage. One must also ensure that they are veiled and that the rubber cover isn't damaged or broken. Once the leads are placed back into the ports, the connection between the jacks and the probes should feel dependable and trustworthy.

When employing test probes, be careful not to contact any exposed metal pieces on them. A lot of contemporary units come with veiled leads that just have a modest metal exposure instead of being like a huge needle. Make sure the leads are adequately insulated and aren't damaged.

One should also never attempt to replace broken test leads. They are not planned for repairs and should be thrown aside as soon as they become broken.

How to Avoid Shocks?

Here are some of the tips you should follow to prevent unwanted shocks:

The first thing you should always assume is that every electrical component inside a circuit is activated. A lot of individuals fail to comply with this guideline and a huge proportion of them endure the repercussions.

Now, the damage doesn't need to be significant to screw up your results. It just takes a minor blunder, and things will probably go poorly from there. That's why you must keep focused and think about every area of your project or else things could not go as planned.

A shock might occur if your body becomes a component of the circuit. It's a rather straightforward idea, however, many people seem to forget about it. Shocks don't always have to be lethal; it relies on several aspects such as the strength of the current, length of time the body spent as a part of
 the circuit, the region of exposed collision, and many others.

The Equipment

As you may know, safety equipment is one of the major things you should be worried about. In other words, one should constantly use protective gloves, hats, insulated mats, and other on-body and near-body equipment. If you're ever exposed to an electrical circuit larger than 50V - wearing safety equipment is necessary.

One should also never work alone when it comes to fiddling with electricity. Having a companion at your side is a fantastic benefit since they can respond accordingly if anything goes wrong. More often than not, patients can't be treated correctly owing to their being alone and unable to ask for aid swiftly.

Avoid working in humid places. While it may not be an obvious thread, humidity may have a significant influence both on your readings and your safety. As we all know, water and electricity don't mix well together and such scenarios should be avoided at all times. regardless of whether you're in a rush and can't wait for another occasion, please refrain from applying multimeters in wet environments.

Understanding the CAT ratings

Every unit bears a certain CAT rating. You need to know what these labels represent to pick the proper tool for the occasion. The basic rule of thumb here is to always select the instrument with the highest category rating in which you are planning on employing it.

A CAT rating refers to the greatest surge of energy a certain model can sustain. In general, the closer you are to the source, the rating will be greater. so, outside conductors are classed as CAT IV because they are at risk of being hit by lightning and so send a significant surge of energy through.

Useful Safety Advice

When measuring resistance, make sure you eliminate the component from the circuit or else you'll obtain incorrect results. If it's not feasible to pull the component out, check sure that the capacitors in the circuit are drained out. Be extremely cautious near capacitors since they may contain a substantial charge and potentially inflict some damage, depending on the total capacity ratings.

If your device doesn't support auto-ranging, make sure you discover and pick the right range manually. In case you don't know how to do it, refer to the user handbook. However, if you pick a lesser range, the unit will show a value of 1. That signifies that the reading is out of the set range and should be fine-tuned appropriately.

Inserting the leads into an incorrect port might lead to a short circuit inside the multimeter, particularly if you measure voltage. It is of paramount significance that you set up everything properly before taking measurements since most mistakes lead to undesirable outcomes that may occasionally be deadly.

When working with AC, ensure the unit is chosen for AC measurements. While it doesn't seem like a major omission, an error in choosing the proper settings may lead to significant effects, particularly for the unit itself.

When you are satisfied with the readings, flip the knob to the OFF position. If the model doesn't have an OFF setting, turn the knob to the highest number on the scale. This is done so that the following person doesn't endanger the integrity of the multimeter by measuring high values while utilizing a sensitive (small) scale.

Worst Case Scenarios

These are some of the troubles that may happen to a person who didn't take all precautionary steps ahead to use a multimeter. Bear in mind, these are all hypothetical scenarios and most of them are, as the header says, a worst-case scenarios. However, to prevent such things, one must first be aware of them.

A lightning strike might induce a transient on the main power line. That event causes an electrical arc between the terminals inside the multimeter itself. As we have indicated, the most essential factor here is the CAT rating.

In case one utilizes a unit with an inappropriate rating, the circuit inside the multimeter might not succeed and even break beyond repair.

The aforementioned arc may reach astounding temperatures of around 5000 degrees Celsius. If one responds normally and pulls the probes out of the source, the arc might be drawn as well and inflict catastrophic harm to the user.

As you can observe, it's of vital significance that you know the qualities of every CAT rating offered so that you can prevent bad problems.

The Conclusion

More often than not, individuals just chose to overlook several safety preventive measures. It's particularly typical in scenarios when working with low voltages, amperage, and whatnot. Needless to add, the low readings don't always suggest low dangers. It's not something that expands proportionately by any means. Instead, one should constantly be aware of the threats as it helps them keep focused and cautious.

There are numerous models available on the market nowadays. Anyone who advises one as the greatest is mistaken since it does need a lot more than just the aesthetic appeal and auxiliary features. Things like ratings, scale sensitivity, design, dependability, and trustworthiness, are also extremely significant and should be treated as relevant.

Don't purchase the first model you come onto, regardless of how costly and cool-looking it is. Instead, take the time to verify its features, CAT rating, quality of the probes, and more. Don't hurry the purchase since the less you care about the items behind the hood, the more you raise the danger of anything going wrong.